Learn
Calculus Quickly

The Complete Guide To Easily Master Calculus
In 100 Solved Equations!

By Math Wizo

First, Get My FREE GIFT!

For a limited time, you can download this book for FREE!
Get it by going to: https://go.mathwizo.com/2

ISBN: 9781796605600

TABLE OF CONTENTS

Calculus 101:
A Walk Through The Basics of Calculus..5

Chapter 1: Limits and Linear Functions.......................................10

Chapter 2: Differential Calculus 28

Chapter 3: Integral Calculus 48

Chapter 4: Solving Some Tough Calculus Problems:
Nesting Integrals, Double Integrals and Triple Integrals 62

CALCULUS 101:
A WALK THROUGH THE BASICS OF CALCULUS

If you are reading this book, you likely want to know two basic things: 1) What is calculus anyway? 2) Does it exist anywhere outside of the mathematician's mind? In other words, does calculus have real-world applications? This book will attempt to answer both questions. Probably, the best way to do this would be to explain why we use calculus anyway in the real world. In fact, calculus plays a big role in many real-life applications.

Before learning calculus, you probably know that there are basic formulations for certain things. For example, you know that the area of a rectangle is the length times the height. This is the formula: A = xy, where A is area, x is the length, and y is the height. For this kind of problem, you don't need calculus. It's simple mathematics. The same holds true of the volume of a box. The volume is the length x the width x the height or V = (length)(width)(height). You don't need calculus for this either.

But what if you have something like this?

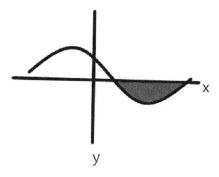

What is the area of the highlighted region? Now that's a lot tougher to do without calculus. The problem is that there are a lot of things in the real world that are not as simple as a rectangle or box, which is where calculus comes in. There are calculus formulas and methods that can be used to determine the answers.

Here's another real-life situation. Suppose you are looking at the rate of cancer in a certain population and you want to know the rate of change at a specific point on the line? You don't need calculus if the curve looks like this:

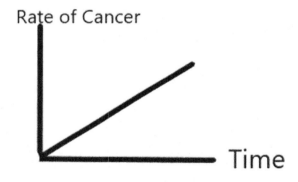

The rate of cancer is the slope of the line, which is the change in y divided by the change in x, where y is the rate of cancer and x is the time. You just need to get two points along the line in the y direction and two corresponding points along the line in the x direction and you have the rate of cancer. Unfortunately, it isn't as simple as this. If you wanted to know the rate of cancer at a specific x point in time and the rate of cancer looked like this next curve, you'd have a problem if you didn't know calculus: .

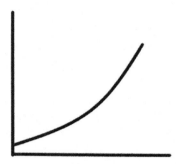

The first question is: Can this even be done? Can you know the rate of change in cancer or the slope of the line as a specific point in time, especially when the curve is not straight? What you'd need to know is the change in y (the rate of cancer) when x (the time) approaches zero, meaning you want to know the instantaneous slope of a curved line. It turns out that, with calculus, this can be done.

So, the answer to the second question posed at the beginning of this book as to whether calculus applies to anything in real life, is yes. Calculus doesn't exist as a fictional thing in a mathematician's mind and, one could argue, it has more real-world applications than figuring out the area of a rectangle as few things in life follow this sort of simple arrangement.

Calculus comes in two varieties, both of which we've already discussed! There is **differential calculus**, which attempts to determine how things change from one moment to the next, such as what the slope of the cancer incidence is at a given point in time on a curve, and **integral calculus**, which joins these small points together in order to determine things like the area of an area under a curve (as in the first example, where we wanted to know the area of highlighted curved region). These things have crucial applications in medicine, biology, engineering, physics, astronomy, and many other areas of life where there isn't a simple geometric arrangement of things.

Calculus deals with **functions** and sometimes nonfunctions. What's a function anyway? Well, a function is simply a term used to describe the relationship between two things. Think of a function as a simple computer. In a simple computer, you can put in a value (the input value) and the computer does a specific function inside the "guts" of the computer to give you an answer (the output value). A nonfunction is a relationship that, when graphed, fails the vertical line test.

The input value of a function is called the independent variable because it allows you to "independently" put in any number you want. In calculus, the independent variable is also called the "**argument**". The output value is the dependent variable because it is "dependent" on the function you decided the computer to do in the first place.

Let's do an example: Say you programmed the computer to do this function: y = x² only in calculus, you write it this way: $f(x) = x^2$ meaning that the function of x is the square of x. Now, just put in any independent variable you want to get the answer. If $x = 2$, then $f(2) = 4$ because $2^2 = 4$. Now let's put in a different value, say 4. This means that $f(4) = 16$. The value is 16.

Now of course, it isn't always that simple: Let's say that the function is this: $f(x) = 2x^2 + 1$. This could be what the slope of a curve represents or any other type of real-world situation. In such cases, you can put in your argument where x = 3. Let's do this problem:
$$f(x) = 2x^2 + 1$$
$$f(3) = 2(3)^2 + 1$$
$$f(x) = 2(9) + 1$$
$$f(x) = 19$$

There are some special functions to think about when you study calculus. These include the following:

- **Constant function**: This is a function that states that no matter what you put in as the argument (independent variable), the value (dependent variable) will always be a constant number. What this looks like is this: $f(x) = c$.
- **Identity function**: This is the function that states that whatever argument you put in, you get a value equal to the argument at all times. It looks like this: $f(x) = x$.

- **Linear function**: This means there is a linear relationship between the argument and the value. It looks like this: $f(x) = mx + b$. This is a "first degree" polynomial function.
- **Quadratic function**: This is more complex but basically uses a "second degree" polynomial, where the graph ends up being a parabola or a curve. It looks like this: $f(x) = ax^2 + bx + c$. For those of you that don't know what a parabola looks like, this is an example:

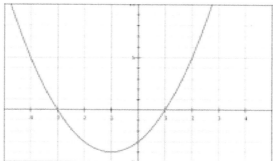

Notice that, in a parabolic or quadratic equation, there are two values of x for every y value, except at the vertex. This is why quadratic equations have two different values of x when they are calculated.

- **Signum function**: This sets up some parameters within the function so that different things are done depending on where the value of x lies. It looks like this:

$$f(x) = \begin{pmatrix} x^2 & x > 0 \\ 0 & x \leq 0 \end{pmatrix}$$

So, if x is greater than zero, the value of the argument is one thing but, if it's less than or equal to zero, the value of the argument is 0. This type of function is also called a piecewise function.

Okay, so these are the basics and there's a lot more to it than that. This can get extremely complicated but we'll make it relatively simple at first. As new things come up, we'll be sure to discuss them before we get too far into it. Let's get started!

CHAPTER 1:
LIMITS AND LINEAR FUNCTIONS

First, we'll talk about limits. Mathematicians use limits when they can't get to the answer to a problem directly. Let's say we have this function: $f(x) = \dfrac{x^2 - 1}{x - 1}$ and let's try to solve for $x = 1$. We get $f(x) = \dfrac{0}{0}$. This just doesn't work because you can't divide anything by zero. This is called an indeterminate value. So, what's done instead is that we try to get as close to one as we can without actually reaching 1. We can plug in numbers close to 1 to see what we can get:

$$f(x) = \frac{x^2 - 1}{x - 1}$$

Now, putting in 0.99 instead of 1, we get this:

$$f(x) = \frac{0.99^2 - 1}{0.99 - 1}$$ which equals 1.99

If we went out to 0.9999999 and plugged it in, we'd get even closer to 2.

So, in mathematics, we can say that as we approach the x-value of 1, we get a limit of 2. How it's written is like this:

$$f(1) = \lim_{x \to 1} \frac{x^2 - 1}{x - 1} = 2$$
.

We can also define the function as a limit: $f(x) = \lim_{x \to \infty} \dfrac{1}{x}$ Well, if we get to infinity with this one, we get an extremely small number, basically zero. This means that the answer to this one is zero.

Let's solve some easy ones:

Problem 1. $f(4) = \lim\limits_{x \to 4} \dfrac{x^2 + 4}{2 + x}$ In this case, we defined the function $f(x)$ as a limit. Basically, you just plug in the value of x as it approaches 4. This direct substitution gives us this:

$$f(x) = \frac{4^2 + 4}{2 + 4}$$ which leads to: $f(x) = \frac{10}{3}$

Problem 2. $f(5) = \lim\limits_{x \to 5} \dfrac{x^2 - 2x + 1}{2x}$ Again, evaluate the limit function $f(x)$ using direct substitution. $f(5) = \dfrac{5^2 - 2(5) + 1}{2(5)}$ or $f(5) = \dfrac{16}{10}$ or $f(x) = \dfrac{8}{5}$.

Problem 3. $f(3) = \lim\limits_{x \to 3} \dfrac{x^2 - 9}{x - 3}$. This looks hard because it gives us $\dfrac{0}{0}$, which is an indeterminate value. But what about simplifying the function by factoring it?

$$f(3) = \lim\limits_{x \to 3} \frac{(x + 3)(x - 3)}{x - 3}$$ Now, this limit can be evaluated cancel $x-3$. This leads to a much easier problem: $f(3) = \lim\limits_{x \to 3} (x + 3)$. Evaluate the limit using direct substitution. Put 3 into the function to get $f(3) = 6$.

Problem 4. Evaluate this limit function: $f(3) = \lim\limits_{x \to 3} \dfrac{x^2 - 2x - 3}{x - 3}$. This appears to be another undefined limit. But we CAN factor the top expression to get this:

$$f(3) = \lim\limits_{x \to 3} \frac{(x - 3)(x + 1)}{x - 3}$$

Then cancel the common factor.

$$f(3) = \lim\limits_{x \to 3} (x + 1)$$

$$f(x) = 4$$

Problem 5. Evaluate this limit function: $f\left(\dfrac{1}{2}\right) = \lim\limits_{x \to \frac{1}{2}} \dfrac{2x - 1}{4x^2 - 1}$. As this appears undefined, it's time to factor things out.

$$f\left(\frac{1}{2}\right) = \lim_{x \to \frac{1}{2}} \frac{2x - 1}{(2x - 1)(2x + 1)}$$

$$f\left(\frac{1}{2}\right) = \lim_{x \to \frac{1}{2}} \frac{1}{(2x + 1)}$$

$$f\left(\frac{1}{2}\right) = \lim_{x \to \frac{1}{2}} \frac{1}{(1 + 1)}$$

$$f\left(\frac{1}{2}\right) = \frac{1}{2}$$

Problem 6. Evaluate this limit function: $f(4) = \lim\limits_{x \to 4} \dfrac{x^2 + x - 20}{x - 4}$. Guess what? It's time to factor again:

$$f(4) = \lim_{x \to 4} \frac{(x - 4)(x + 5)}{x - 4}$$ Now we can divide by $x-4$.

$$f(4) = \lim_{x \to 4} (x + 5)$$ (now solve for the limit by putting 4 into the equation).

$$f(4) = 9$$

Problem 7. Evaluate this limit function: $f(-6) = \lim\limits_{x \to -6} \dfrac{2x^2 + 13x + 6}{x + 6}$
This will take some creative factoring. Try this:

$$f(-6) = \lim_{x \to -6} \frac{(x + 6)(2x + 1)}{x + 6}$$

Now we can divide by $x + 6$ to get this:

$$f(-6) = \lim_{x \to -6} (2x + 1)$$

$$f(-6) = -11$$

Problem 8. Evaluate this limit function: $f(1.5) = \lim_{x \to 1.5} \dfrac{2x - 3}{6x^2 - 13x + 6}$

When in doubt, see if you can factor the expressions.

$$f(1.5) = \lim_{x \to 1.5} \frac{2x - 3}{(2x - 3)(3x - 2)}$$

Now you can cancel $2x-3$ from the top and the bottom to get:

$$f(1.5) = \lim_{x \to 1.5} \frac{1}{(3x - 2)}$$

This is solvable:

$$f(1.5) = \frac{1}{3(1.5) - 2}$$

$$f(1.5) = 0.4$$

Problem 9. Evaluate this limit function: $f(1) = \lim_{x \to 1} \dfrac{1 - \sqrt{x}}{1 - x}$ First of all, don't be frightened of the square root. This is actually fairly easy. Multiply by the conjugate of the numerator. This leads to this:

$$f(1) = \lim_{x \to 1} \frac{(1 - \sqrt{x})(1 + \sqrt{x})}{(1 - x)(1 + \sqrt{x})}$$

$$f(1) = \lim_{x \to 1} \frac{1 - x}{(1 - x)(1 + \sqrt{x})}$$

$$f(1) = \lim_{x \to 1} \frac{1}{(1 + \sqrt{1})}$$

$$f(1) = \frac{1}{2}$$

$$f\left(\frac{1}{9}\right) = \lim_{x\to\frac{1}{9}} \frac{9x - 1}{3\sqrt{x} - 1}$$

Problem 10. Evaluate this limit function: Again, don't be afraid of the square root. This time, multiply by the conjugate of the denominator:

$$f\left(\frac{1}{9}\right) = \lim_{x\to\frac{1}{9}} \frac{9x - 1}{3\sqrt{x} - 1}$$

$$f\left(\frac{1}{9}\right) = \lim_{x\to\frac{1}{9}} \frac{(9x - 1)(3\sqrt{x} + 1)}{(3\sqrt{x} - 1)(3\sqrt{x} + 1)}$$

$$f\left(\frac{1}{9}\right) = \lim_{x\to\frac{1}{9}} \frac{(9x - 1)(3\sqrt{x} + 1)}{9x - 1}$$

Now cancel the common factor, leading to this:

$$f\left(\frac{1}{9}\right) = \lim_{x\to\frac{1}{9}} \frac{(3\sqrt{x} + 1)}{1}$$

Now solve the equation:

$$f\left(\frac{1}{9}\right) = 3\left(\frac{\sqrt{1}}{\sqrt{9}}\right) + 1$$

$$f\left(\frac{1}{9}\right) = 3\left(\frac{1}{3}\right) + 1$$

$$f\left(\frac{1}{9}\right) = 2$$

Problem 11. Evaluate this limit function: $f(3) = \lim_{x\to3} \frac{2x^2 - 4x - 6}{x - 3}$ Try to factor out $x-3$ from the top as a way to do this:

$$f(3) = \lim_{x\to3} \frac{2(x - 3)(x + 1)}{x - 3}$$

Now you can cancel the $x - 3$ from the top and bottom:

$$f(3) = \lim_{x \to 3} \frac{2(x+1)}{1}$$

This is easily evaluated:

$$f(3) = 2(3+1)$$

$$f(3) = 8$$

Problem 12. Evaluate this limit function: $f(-1) = \lim_{x \to -1} \frac{3x^2 + x - 2}{x+1}$ Again, you need to know how to factor this out to get $x+1$ isolated from the top half of the equation. This leads to this:

$$f(-1) = \lim_{x \to -1} \frac{(3x-2)(x+1)}{x+1}$$

Now, you can cancel the $x+1$ from both the top and bottom of the equation:

$$f(-1) = \lim_{x \to -1} \frac{(3x-2)}{1}$$

Now evaluate the limit:

$$f(-1) = 3(-1) - 2$$

$$f(-1) = -5$$

Problem 13. Evaluate this limit function:
$$f(3) = \lim_{x \to 3} \frac{x-3}{2x^2 - 9x + 9}$$
So, you can try to factor out $x-3$ from the bottom half of the equation. This takes some skill but it can be done. This leaves you with this:

$$f(3) = \lim_{x \to 3} \frac{x-3}{(x-3)(2x-3)}$$

Now, cancel the $x-3$ from the top and bottom of the equation:

$$f(3) = \lim_{x \to 3} \frac{1}{(2x-3)}$$

Now the limit can be evaluated:

$$f(3) = \frac{1}{2(3)-3}$$

$$f(3) = \frac{1}{3}$$

Problem 14. Evaluate this limit function:
$$f(-4) = \lim_{x \to -4} \frac{x+4}{2x^2 + 6x - 8}$$
By now, you should be getting this. You need to factor the bottom of the equation to tease out $x+4$ out of the polynomial. (Incidentally, you can use the quadratic equation to solve the bottom part of the equation.)

$$f(-4) = \lim_{x \to -4} \frac{x+4}{2(x+4)(x-1)}$$

Now divide both the top and bottom by $x+4$, leaving this:

$$f(-4) = \lim_{x \to -4} \frac{1}{2(x-1)}$$

$$f(-4) = \frac{1}{2(4-1)}$$

$$f(-4) = \frac{1}{6}$$

Problem 15. Evaluate this limit function: $f(4) = \lim\limits_{x \to 4} \dfrac{x^2 - 7x + 12}{x - 4}$ Again, this is a lesson that involves both the properties of limits and algebraic factoring. Hopefully, you've already taken algebra and have a good understanding of factoring. Start by factoring $x-4$ out of the top of the equation:

$$f(4) = \lim\limits_{x \to 4} \frac{(x - 4)(x - 3)}{x - 4}$$

Now cancel the $x-4$ from the top and bottom of the equation:

$$f(4) = \lim\limits_{x \to 4} \frac{(x - 3)}{1}$$

$$f(4) = 4 - 3$$

$$f(4) = 1$$

By now you may be asking why limits are so important. It's because there are equations and graphs that actually become difficult to solve without limits. You may need to use this concept of limits to solve harder problems.

Linear Functions

We've already talked about what a linear function is. It's basically something that graphs out when plotted to be a straight line. For example, say you want to know what the slope of a straight line is and you've got this:

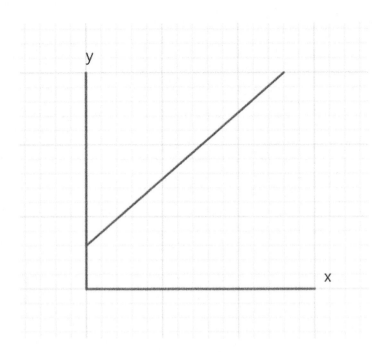

The slope of line can be defined in several different ways. In calculus, the points on a line that determine the slope aren't (x_1, y_1) and (x_2, y_2). Instead, the two points on a line that determine the slope are $(x_1, f(x_1))$, and $(x_2, f(x_2))$. A line can be defined by its slope plus a constant, which is the number on the y-axis when $x = 0$ on the graph, called the y-intercept. In other words, the line can be defined as $y = mx + b$. The value of m is the slope of the line, while b is the constant. In the above example, we can see that $b = 3$ on the graph [because we can count three squares up on the y-axis from the $(0, 0)$ point.] But what is the slope?

The slope can be calculated by using this equation:

$$m = \frac{\Delta y}{\Delta x}$$ in which the triangle or delta sign means "change in". It means also that $$m = \frac{f(x_2) - f(x_1)}{x_2 - x_1}$$

The equation in calculus is not the same as in algebra with $y = mx + b$ but is instead this: $f(x) = mx + b$.

Problem 16. You've got a straight line in which there are two points: $(2, 4)$ and $(4, 16)$. What is the slope and what is the equation that defines the line?

Start with the slope formula:

$$m = \frac{f(x_2) - f(x_1)}{x_2 - x_1}.$$

Now plug the numbers into the equation to get this:

$$m = \frac{16 - 4}{4 - 2}$$

$$m = \frac{12}{2}$$

$$m = 6$$

Now we know that $f(x) = 6x + b$. So what's b? It can be calculated by plugging in one of the points in the equation, such as $(2,4)$. Try this: $4 = 6(2) + b$, $b = 4{-}12 \text{ or } b = -8$. The equation is this:

$$f(x) = 6x - 8.$$

Problem 17. You have a line with two points: $(4, 10)$ and $(8, 4)$. What's the equation?

$$m = \frac{10 - 4}{4 - 8}.$$

$$m = \frac{6}{-4}.$$

$$m = -\frac{3}{2}$$

Now solve for b:

$$f(x) = -\frac{3}{2}x + b.$$

Now plug in one of the points: $(8, 4)$

$$4 = -\frac{3}{2}(8) + b.$$

$$b = 16$$

$$f(x) = -\frac{3}{2}x + 16$$

Problem 18. You have two points $(-1, 4)$ and $(11, 14)$ and you want to know the equation of the line that passes through these two points. First find the slope:

$$m = \frac{14 - 4}{11 - (-1)}$$

$$m = \frac{10}{12}$$

$$m = \frac{5}{6}$$

Now it's time to figure out the y-intercept or the value of b in this equation:

$$f(x) = \frac{5}{6}x + b$$

$$4 = \frac{5}{6}(-1) + b$$

$$b = \frac{24}{6} + \frac{5}{6}$$

$$b = \frac{29}{6}$$

$$f(x) = \frac{5}{6}x + \frac{29}{6}$$

Problem 19. You have two points on a line: $(10, 2)$ and $(4, 1)$ What is the equation of the line that passes through these two points?

$$m = \frac{f(x_2) - f(x_1)}{x_2 - x_1}$$

$$m = \frac{1 - 2}{4 - 10}$$

$$m = \frac{1}{6}$$

$$f(x) = \frac{1}{6}x + b$$

$$1 = \frac{1}{6}4 + b$$

$$b = \frac{1}{3}$$

The equation:

$$f(x) = \frac{1}{6}x + \frac{1}{3}$$

Problem 20. You have a line with two points $(-2, -10)$ and $(-4, 6)$. What is the equation of the line that passes through these two points?

$$m = \frac{f(x_2) - f(x_1)}{x_2 - x_1}$$

$$m = \frac{6 - (-10)}{-4 - (-2)}$$

$$m = \frac{6 + 10}{-4 + 2}$$

$$m = -8$$

Now we'll determine the value of b:

$$f(x) = -8x + b$$

$$6 = -8(-4) + b$$

$$6 = 32 + b$$

$$b = -26$$

The equation:

$$f(x) = -8x - 26$$

Problem 21. Find the equation. (We'll do something else after that). You have a line with two points:

$(-10, 4)$ and $(4, 10)$. Find the equation:

$$m = \frac{f(x_2) - f(x_1)}{x_2 - x_1}$$

$$m = \frac{10 - 4}{4 - (-10)}$$

$$m = \frac{6}{14} = \frac{3}{7}$$

$$f(x) = \frac{3}{7}x + b$$

Now determine the y-intercept: Plug in $(4, 10)$

$$f(x) = \frac{3}{7}x + b$$

$$10 = \frac{3}{7}(4) + b$$

$$b = \frac{70}{7} - \frac{12}{7}$$

$$b = \frac{58}{7}$$

So, the equation is:

$$f(x) = \frac{3}{7}x + \frac{58}{7}$$

Part of calculus involves graphing different equations and determining the slope of lines, even if the line isn't straight. This will be the topic of later chapters when we deal with the slope of a curved line at a specific point on the line. For now, we'll take this opportunity to graph some easy, straight lines, given the linear function.

Problem 22. You have a line with the function: $f(x) = -2x + 4$. What does this look like on a graph?

You start with knowing that the intercept of $f(x) = 4$. This means that one point on the line must be $(0,4)$. You also know that the slope is -2. What you want to know is another point on the line in order to be able to graph the equation. Logically, this would be the situation of $(x,0)$, which is the value of x when $f(x) = 0$. Simply solve the equation with $f(x)$ at 0.

$$0 = -2x + 4.$$

$$-2x = -4$$

$$x = 2$$

This leads to the two points: $(0,4)$ and $(2,0)$

Graph the line using these two points:

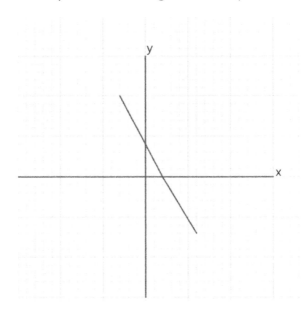

As you can see, the slope is downward or "negative" as is reflected in the graph.

Problem 23. Graph the equation: $f(x) = 4x - 6$ This leads to the b constant being -6 and the known point on the line of this: $(0, -6)$. So, now it's up to determine another point, that being $(x, 0)$. Again, plug in the equation the value of $f(x)$ being 0:

$$0 = 4x - 6$$

$$x = \frac{6}{4} = \frac{3}{2}$$

This leads to two points: $(0, -6)$ and $\left(\frac{3}{2}, 0\right)$

The graph is:

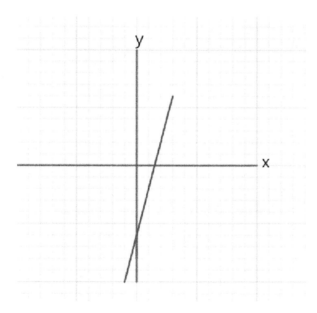

Problem 24. Graph the equation of the line: $f(x) = \frac{1}{2}x + 6$ This leads to the known constant of 6 and the known point of $(0, 6)$. Let's solve for the point $(x, 0)$.

$$0 = \frac{1}{2}x + 6$$

$$x = -(2)6$$

$$x = -12$$

This leads to two points: $(0, 6)$ and $(-12, 0)$. Let's graph the line:

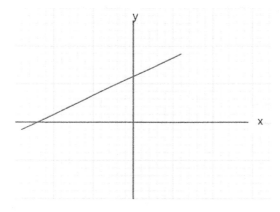

Problem 25. Graph the equation: $f(x) = -3x + 6$

The first thing you know is that $b = 6$ and that one point is $(0, 6)$. Let's solve for $(x, 0)$.

$$0 = -3x + 6$$

$$x = 2$$

Now you have the point $(2, 0)$ along with the point $(0, 6)$, which leads to this graph:

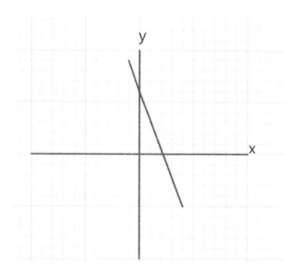

CHAPTER 2:
DIFFERENTIAL CALCULUS

There is a reason we've spent an entire chapter on things like limits and slope. As mentioned, it's one thing to find the slope of a line when the line is straight. It's another thing altogether to find the slope of a line when the curve isn't straight. For one thing, a curve has a different slope throughout its length, and depending on the location on the curve, the slope could be very steep indeed. What you want to know is the slope of a curve at a specific point on the graph. This is referred to not as "slope" but as the "derivative" because it does not, practically speaking, have two points with which to calculate the actual slope.

In fact, when you think of slope, which is slope = the derivative = $\frac{\Delta y}{\Delta x} = \frac{(f(x + \Delta x) - f(x)}{\Delta x}$. What this looks like is shown here:

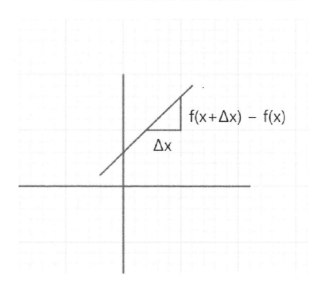

$$f(x+\Delta x) - f(x)$$

$$\Delta x$$

When you have a straight line, the slope is given easily by knowing two points. But, with a curve, you can still calculate the slope at a specific point by using the tangent of the curve at that specific point. The tangent is a line that represents the slope at a particular point along the curve and the slope looks like this:

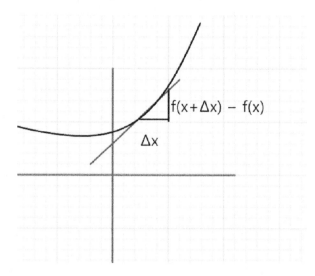

This is where limits come in. The slope at a specific point is found when Δx is an extremely small number if the goal is to get the slope at that specific point. The trick to doing this? It relates to limits because what you're calculating the slope as the $\lim\limits_{\Delta x \to 0} slope$. That gets you as close to determining the exact slope of a curve at a specific point as possible (mathematically speaking). Mathematicians cheat reality all the time through processes like limits that get an answer when it doesn't seem impossible that an answer can be found.

Let's look at determining the derivative (slope) for the equation: $f(x) = x^2$. The graph of this works out to be a parabola so the slope differs over the value of x. The graph would look something like this:

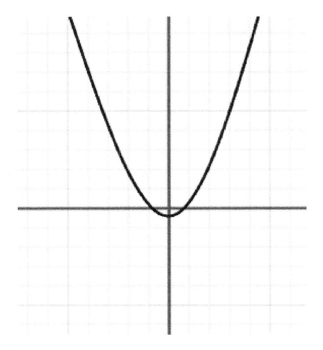

If $f(x) = x^2$, then this is also true: $f(x + \Delta x) = (x + \Delta x)^2$. Got it?

Now expand this equation out so you can see it in quadratic form:

$$f(x + \Delta x) = x^2 + 2x(\Delta x) + (\Delta x)^2$$

So far, no tricks, just some algebraic expansion. Now we'll calculate the slope of this equation:

$$slope = \frac{\Delta y}{\Delta x} = \frac{(f(x + \Delta x) - f(x))}{\Delta x}$$

The slope then is:

$$m = \frac{x^2 + 2x\Delta x + (\Delta x)^2 - x^2}{\Delta x}$$

In the above equation, we took the expanded $f(x + \Delta x)$ and put it into the slope equation.

Now we can simplify it:

$$m = \frac{2x\Delta x + (\Delta x)^2}{\Delta x}$$

And we can simplify it even more by dividing by Δx in the top and bottom of the equation:

$$m = 2x + \Delta x$$

So, when thinking of a slope at a given point on the line, we're looking at a sort of limit situation, such as this:

$$\lim_{\Delta x \to 0} (2x + \Delta x)$$

This means that the slope at this point is this:

$$m = 2x \text{ (Yes, it's really that simple!)}$$

So, in calculus, the slope at a given point on a line or curve is called the derivative. Instead of always using the term "Δx heads toward zero", mathematicians who study calculus call this dx. In addition, the derivative notation is written as $\frac{d}{dx}$. So, when saying the derivative of something you just write:

$$\frac{d}{dx}x^2 = 2x$$
("the derivative of x² equals 2x")

To make things even more complicated, there is still another way to write the derivative. This involves this notation:

$$f(x) = x^2 \text{ so } f'(x) = \frac{d}{dx}x^2 = 2x$$

So, basically $f'(x)$ is another way of saying "the derivative of $f(x)$". As you will see, equations will often have more than one complex variable, often "nested" into one another. For example, you can have $f(g(x))$ in which there are functions and two derivatives that can be calculated. This happens when you have a complex function that requires you to "substitute" a variable expression into a complex set of variables that are too complicated to do otherwise. The rule for this, as you'll see is called the "chain rule".

To make it simple for you, mathematicians have worked out some "rules" for derivatives that it would be worthwhile to know. These include the following:

- The derivative of a constant value (like just a number) is always 0.
- The derivative of x is always 1.
- The derivative (slope) of a line (such as $y = 3x + 3$) is always the value of m, such as 3 in this case.
- The derivative of x^2 is $2x$.
- The derivative of a square root or $\sqrt{x}$ is $\dfrac{1}{2\sqrt{x}}$
- The derivative of x^n is nx^{n-1}
- The derivative of $\dfrac{1}{f}$ is $-\dfrac{f'}{f^2}$
- The derivative of $\dfrac{1}{x}$ is $-\dfrac{1}{x^2}$
- The derivative of the constant e to the power of x, e^x is e^x
- The derivative of a function multiplied by a constant cf is cf'
- In the sum of two functions $f + g$, the derivative is $f' + g'$ (so basically, you can add derivatives together)
- In the difference of two functions $f-g$, the derivative is $f' - g'$ (so basically, you can subtract derivatives)
- In the product of two functions $f(x) \cdot g(x)$ multiplied together, the product rule for the derivative is $f(x) \cdot g'(x) + f'(x) \cdot g(x)$
- In two functions are divided as a quotient, $\dfrac{f}{g}$, the quotient rule for derivatives is $\dfrac{f'g - g'f}{g^2}$
- In the chain rule, for $f(g(x))$, the derivative is $f'(g(x))(g'(x))$.
- In trigonometry, when x is in radians: the derivative of $sin(x)$ is $cos(x)$
- When x is in radians: the derivative of $cos(x)$ is $-sin(x)$
- When x is in radians: the derivative of $tan(x)$ is $sec^2(x)$

- When x is in radians, the derivative of $cotx$ is $-csc^2(x)$
- When x is in radians, the derivative of $secx$ is $secxtanx$
- When x is in radians, the derivative of $cscx$ is $-cscxcotx$

Let's look at some derivative questions:

Problem 26. Let's solve something more complicated first so you can actually believe that the rules listed above actually work. Calculate this derivative the hard way: What is the derivative of $f(x) = x^3$?

First you need to calculate $f(x + \Delta x)$ or $f(x + dx)$, which in this case is $(x + dx)^3$

Expand this out: $x^3 + 3x^2dx + 3xdx^2 + dx^3$

The slope formula is this:
$$f'(x) = \frac{x^3 + 3x^2dx + 3xdx^2 + dx^3 - x^3}{dx}$$
or simplified, you get this:

$$f'(x) = \frac{3x^2dx + 3xdx^2 + dx^3}{dx}$$

Now cancel a dx in the top and bottom by dx and you get:

$$f'(x) = \frac{3x^2 + 3xdx + dx^2}{1}$$

Now, what happens when we take dx or Δx and make it approach zero? Whoosh, you get rid of dx!!

If $f(x) = x^3$, then $f'(x) = 3x^2$

So, are you a believer now in the idea that the derivative of any exponent x^n is nx^{n-1}?

Problem 27. What is $f'(x)$ if $f(x) = 3x^3 + 2x^2$?

Now, we could calculate all the derivatives like we did in problem 26 but, hopefully, we've proven that this is not necessary so we can calculate the derivative in the two parts separately:

If $g(x) = 3x^3$ then $g(x) = 3(3)x^2 = 9x^2$

And if $h(x) = 2x^2$ then $h'(x) = 3(2x) = 6x$

Then we know that, in cases of $g(x) + h(x)$, then $g'(x) + h'(x)$ gives the derivative of the function as this:

If $f(x) = 3x^3 + 2x^2$ then $f'(x) = 9x^2 + 6x$ Again, this isn't a magic trick, you've worked it out the hard way in order to prove that the derivative formulas above work. It's a matter of figuring out the slope of a line when the line isn't really a line after all and the slope is the slope at a specific point because dx or Δx in this case is really, really close to zero. It's just calculus.

Problem 28. What is the derivative of $f(x)$ if $f(x) = 5x^2 + x^3 - 7x^4$? Not so hard now, is it?

If $g(x) = 5x^2$, then $g'(x) = 5(2x) = 10x$

If $h(x) = x^3$, then $h'(x) = 3x^2$

If $k(x) = -7x^4$, then $k'(x) = -7(4)x^3 = -28x^3$

$$f'(x) = 10x + 3x^2 - 28x^3$$

Problem 29. Find $f'(x)$ if $f(x) = 2x^3 - 3x^4 + \dfrac{1}{x}$.

Again, separate it out:

If $g(x) = 2x^3$, then $g'(x) = 2(3)x^2 = 6x^2$

If $h(x) = -3x^4$, then $h(x) = -3(4)x^3 = -12x^3$

If $k(x) = \dfrac{1}{x}$, then $k'(x) = \dfrac{-1}{x^2}$

The derivative then is:

$$f'(x) = 6x^2 - 12x^3 - \frac{1}{x^2}$$

Problem 30. What is the derivative of $f(x)$ if $f(x) = x^3 - 5x^2 + 3x$? This should be old hat by now:

If $g(x) = x^3$, then $g'(x) = 3x^2$

If $h(x) = -5x^2$, then $h'(x) = -5(2)x = -10x$

If $k(x) = 3x$, then $k'(x) = 3$

$$f'(x) = 3x^2 - 10x + 3$$

Problem 31. Find the derivative of $f(x)$ if $f(x) = (x-1)(x+3)$.

It's probably best to multiply it out to get this: $f(x) = x^2 + 2x - 3$

$(x^2)' = 2x$

$$(2x)' = 2$$

Putting it altogether, you get:

$$[(x-1)(x+3)]' = 2x + 2$$

Problem 32. What is the derivative of $f(x)$ if $(x) = \left(\dfrac{3x^3 - x^2}{x}\right)$?

To start with, just divide the top and bottom by x to get this:

$$f(x) = \left(\frac{3x^2 - x}{1}\right)$$

By now, this should be easy:

$$(3x^2 - x)' = 6x - 1$$

Problem 33. What is the derivative of $f(x)$ if $f(x) = x\sin(x)$? Okay, so this is the product rule (and a little bit of trigonometry but as long as you use the rules, it should be easy). When multiplying two things and getting their functions, the rule is this: The derivative of $f(x)g(x)$ is $f'(x)(g(x)) + f(x)(g'(x))$. In this case, you've got $f(x) = x$, $g(x) = \sin(x)$, $f'(x) = 1$ and $g'(x) = cosx$. Plugging these things in you get this:

$$[f(x)g(x)]' = (1)(\sin(x)) + x(\cos(x))$$

Simplifying completely:

$[f(x)g(x)]' = \sin(x) + x\cos(x)$

Problem 34. Find $h'(x)$ if $h(x) = \dfrac{\cos(x)}{x}$

This involves the quotient rule. Pretend that you have two separate functions to deal with and use the quotient rule: The top part is $f(x) = \cos(x)$ and $f'(x) =- \sin(x)$. The bottom part is $g(x) = x$ and $g'(x) = 1$. The quotient rule is: $\dfrac{d}{dx}\left(\dfrac{f(x)}{g(x)}\right) = \dfrac{f'(x)g(x) - g'(x)f(x)}{(g(x))^2}$. Just plug in the values: $\left(\dfrac{-\sin(x)x - 1(\cos(x))}{x^2}\right)' = \dfrac{-x\cos(x) - \sin(x)}{x^2}$

Problem 35. Find the derivative of $h(x)$ if

$$h(x) = \frac{1}{x^2 + 1}$$

Again, here you have a function that is a quotient, so use the quotient rule to find the derivative:

$f(x) = 1$ and $f'(x) = 0$, while $g(x) = x^2 + 1$ and $g'(x) = 2x$. Now substitute and divide it out:

$$\frac{f'(x)g(x) - g'(x)f(x)}{(g(x))^2}$$

$$\frac{(0)(x^2 + 1) - (2x)(1)}{(x^2 + 1)^2}$$

$$h'(x) = \frac{-2x}{(x^2 + 1)^2}$$

Problem 36. Find: $\frac{d}{dx}((5x - 2)^3)$. Now you can multiply all of this out but you don't have to if you use the chain rule: In the chain rule, where $f(g(x))$, the derivative is $(f'(g(x))g'(x)$

In this case: $(5x - 2)^3$ is made of (g^3) and $g = (5x-2)$ so substituting " $5x-2$" with "g":

$$f(g) = g^3 \text{ and}$$

$g(x) = 5x - 2$

Now, do the derivatives:

$$f'(g) = 3g^2 \text{ and}$$

$$g'(x) = 5$$

Now multiply it out and putting back in place of g:

$$\frac{d}{dx}((5x - 2)^3) = 3(5x - 2)^2(5) = 15(5x - 2)^2$$

Problem 37. Find the derivative of : $f(x) = e^{2x^2 + x}$. This looks really hard but isn't when you use the chain rule:

In this case, you can divide it into four parts: $f(x) = e^{g(x)}$ where g(x) is $2x^2 + x$

$f'(x) = e^{g(x)} \cdot g'(x)$

$$g(x) = 2x^2 + x$$

$$g'(x) = 4x + 1$$

If we have $f(g(x))$, the derivative is $f'(g(x))(g'(x)$

Plugging it all in you get: $f'(x) = (4x + 1)e^{(2x^2 + x)}$

$$f'(x) = (4x + 1)e^{(2x^2 + x)}$$

Anyway, you get the idea. This chain rule is complicated and there just isn't any easy way to explain it. When it gets complicated, you can substitute a second variable in for the complicated ones.

Problem 38. What is the derivative of $f(x) = (2x^3 + 1)(3x + 2)$? In other words, what is the answer to this?

$$f'(x) = \frac{d}{dx}(2x^3 + 1)(3x + 2)$$

This involves the product rule: In two functions $f(x) \cdot g(x)$ multiplied together, the derivative is $f(x)g'(x) + f'(x)g(x)$:

$$f(x) = 2x^3 + 1$$

$$f'(x) = 6x^2$$

$$g(x) = 3x + 2$$

$$g'(x) = 3$$

The derivative is: $(2x^3 + 1)(3) + 6x^2(3x + 2)$

Simplify: $f'(x) = 6x^3 + 3 + 18x^3 + 12x^2$

$$f'(x) = 24x^3 + 12x^2 + 3$$

Question 39. Find the derivative of: $f(x) = \dfrac{2x^3}{2x + 2}$

Again, the function is the quotient of two different functions. Use the quotient rule:

$$g(x) = 2x^3 \ so \ g'(x) = 6x^2$$

$$h(x) = 2x + 2 \ so \ h'(x) = 2$$

The derivative is formula:

$$\frac{g'(x)h(x) - h'(x)g(x)}{(h(x))^2}$$

$$f'(x) = \frac{(6x^2)(2x + 2) - 2(2x^3)}{(2x + 2)^2}$$

$$f'(x) = \frac{8x^3 + 12x^2}{(2x + 2)^2}$$

Problem 40. Find the derivative: $f(x) = e^{(x^3 + 2)}$.

This is a chain rule problem again. In it, you can break it down to this:

$$f(x) = e^{g(x)} \text{ (where } g(x) = x^3 + 2)$$

$f'(x) = (g'(x))e^{g(x)}$

$g(x) = x^3 + 2$

$g'(x) = 3x^2$

Remember that, with the chain rule: If we are dealing with $f(g(x))$, the derivative is $(f'(g(x)))(g'(x))$ so the derivative is this:

Resubstituting $x^3 + 2$ for g: $(e^{(x^3 + 2)})(3x^2)$

$$\frac{dy}{dx}(f(x)) = 3x^2 e^{(x^3 + 2)}$$

Now we'll get into the topic of second derivatives. This is basically the derivative of a derivative, referred to as $f''(x)$ or referred to as $\frac{d^2y}{dx^2}$. So, what does it mean? Well, the first derivative is the slope or the "rate of change" of a line, then the second derivative tells you whether or not the slope is increasing or decreasing, and at what rate.

A real-world example is that of distance, speed, and acceleration. If you're driving a car over a certain distance y, you're going at a certain velocity, such as 60 miles per hour or dy/dx. Let's say you're accelerating at a certain rate, such as 1 mile/hr^2. Acceleration is the second derivative of the motion function and the first derivative is the velocity.

Now that you know derivatives, you can calculate the second derivative.

Problem 41. What is the second derivative of $f(x) = 3x^3 - 4x^2 + x$?

$$f(x) = 3x^3 - 4x^2 + x$$

$$f'(x) = 9x^2 - 8x + 1$$

$$f''(x) = 18x - 8$$

Problem 42. What is the second derivative of $f(x) = 4x^4 + 5x^3 - 2x + 7$?

$$f(x) = 4x^4 + 5x^3 - 2x + 7$$

$f'(x) = 16x^3 + 15x^2 - 2$ (This is the first derivative).

$$f''(x) = 48x^2 + 30x$$

Problem 43. What is the second derivative of $x^2\sin(x)$?

This involves the product rule: $f(x) = x^2$ and $f'(x) = 2x$

$g(x) = \sin(x)$ and $g'(x) = \cos(x)$

Remember the product rule: $\dfrac{d}{dx}[f(x) \cdot g(x)] = f(x)g'(x) + f'(x)g(x)$ so

$\dfrac{d}{dx}(x^2\sin(x)) = x^2\cos(x) + 2x(\sin(x))$

The first derivative is: $x^2\cos(x) + 2x\sin(x)$ This expression contains two products of two functions.

To find the second derivative split the first up into the first half and the second half adding the two parts in the end.

$f'(x)$ for the first half is x^2 and $f''(x) = 2x$

$g'(x)$ for the first half is $\cos(x)$ and $g''(x) = -\sin(x)$

The second derivative for the first half is: $2x\cos(x) + x^2(-\sin(x))$

Now find the derivative of the second half, repeat the process:

$f'(x)$ for the second half is $2x$ and the $f''(x) = 2$

$g'(x)$ for the second half is $\sin(x)$ and the $g''(x) = \cos(x)$

The second derivative for the second half is: $2x\cos(x) + 2\sin(x)$

Now put both halves together: $2x\cos(x) - x^2\sin(x) + 2x\cos(x) + 2\sin(x)$ and combine like terms.

Finally, the second derivative is: $4x\cos(x) - x^2\sin(x) + 2\sin(x)$

Problem 44. What is the second derivative of 1/x?

According to the rules, the first derivative of $1/x$ is -1/x²

To find the second derivative, rewrite the first derivative as
$f'(x) = -x^{-2}$. The use the power rule that says $\dfrac{d}{dx}(x^n) = nx^{n-1}$.

Therefore, the second derivative is $(-1)(-2)x^{-2-1} = 2x^{-3} = \dfrac{2}{x^3}$

Problem 45. What is the second derivative of $f(x) = \dfrac{x + x^2}{2x + 1}$?

Start with the quotient rule to find the first derivative: $g(x) = x + x^2$ (the top half of the quotient)

$$h(x) = 2x + 1 \text{ (the bottom half of the quotient)}$$

$g'(x) = 1 + 2x$

$h'(x) = 2$

Now, get the first derivative:

$$\frac{(1 + 2x)(2x + 1) - (2)(x + x^2)}{(2x + 1)^2}$$

$$\frac{(4x^2 + 4x + 1) - (2x + 2x^2)}{(4x^2 + 4x + 1)}$$

$$\frac{(2x^2 + 2x + 1)}{(4x^2 + 4x + 1)}$$

Now, we get to the even messier second derivative:

Split it up: $g''(x) = 4x + 2$ (the derivative of the top)

$h''(x) = 8x + 4$ (the derivative of the bottom)

Now use the quotient rule:

$$\frac{f'(x)g(x) - g'(x)f(x)}{(g(x))^2}$$

$$\frac{(4x + 2)(4x^2 + 4x + 1) - (8x + 4)(2x^2 + 2x + 1)}{(4x^2 + 4x + 1)^2}$$

Okay, let's try to simplify this:

$$\frac{16x^3 + 24x^2 + 12x + 2 - 16x^3 - 24x^2 - 16x - 4}{(4x^2 + 4x + 1)^2}$$

$$f''(x) = \frac{-4x - 2}{(4x^2 + 4x + 1)^2}$$

Next, we'll get into the topic of partial derivatives. What is a partial derivative? Well, imagine you had an equation that had two variables x and y? Can you determine the derivative of this sort of thing? As it turns out, you can! It's actually very simple. You can't determine the derivative of both variables at once but you can treat one as a constant, differentiating for the other. It's called "holding the variable constant". The function notation for this type of function is $f(x,y)$ and the notation for the derivative with respect to x is f_x or $\frac{df}{dx}$. The notation for the derivative with respect to y is f_y or $\frac{df}{dy}$. Remember, as we differentiate, we treat each variable, one at a time, as a constant, but we cannot assign a constant value to them. Let's do a few:

Problem 46. Define the derivative of x and y with respect to each other if the equation is: $f(x,y) = 3x^4 + 2y^2$.

This is easy: Hold y constant so the derivative of the y-term is 0 The derivative with respect to x is: $3x^4 + 0$ so you get: $\frac{df}{dx} = 12x^3$

Now, do the same thing for y: $\frac{df}{dy} = 0 + 4y$ so $\frac{df}{dy} = 4y$

Problem 47. What is the derivative of x and y with respect to the other if the equation is $f(x,y) = \dfrac{x}{y^2}$?

Let each variable be a constant as we differentiate.

This too is easy: Rewrite the function as $f(x,y) = \dfrac{1}{y^2}x.$ Then $\dfrac{df}{dx} = \dfrac{d}{dx}\left(\dfrac{1}{y^2}x\right) = \dfrac{1}{y^2}$

First, rewrite $f(x,y)$ so we can differentiate using the power rule: $f(x,y) = xy^{-2}$ so $\dfrac{df}{dy} = -2xy^{-3} = -\dfrac{2x}{y^3}.$

Problem 48. Find the second derivative of x and y with respect to each other if the equation is this:

$$f(x,y) = xy^3 + 2x + 3xy$$

Start with the derivative with respect to x and write it treating y as a constant:

First derivative with respect to x is: $\dfrac{df}{dx} = y^3 + 2 + 3y$

The second derivative with respect to x is $\dfrac{d^2f}{dx^2} = 0$

Continue with the first derivative with respect to y and write it treating x as a constant:

First derivative with respect to y is: $\dfrac{df}{dy} = 3xy^2 + 3x$

The second derivative with respect to y is $\dfrac{d^2f}{dx^2} = 6xy$

Problem 49. Find the derivative of x and y with respect to one another if the equation is:

$$f(x,y) = \frac{xy^2 + 1}{y} + xy$$

Rewrite the equation as $f(x,y) = \dfrac{xy^2}{y} + \dfrac{1}{y} + xy \text{ or } xy + \dfrac{1}{y} + xy$

First treat y as a constant and differentiate with respect to x. The derivative with respect to x is: $\dfrac{df}{dx} = \dfrac{y^2}{y} + y = y + 0 + y = 2y$

Rearrange $f(x,y)$ into $f(x,y) = \dfrac{xy^2}{y} + \dfrac{1}{y} + xy = xy + \dfrac{1}{y} + xy = 2xy + \dfrac{1}{y}$

Next, treat x as a constant and differentiate with respect to y:

$$\frac{df}{dy} = 2x - \frac{1}{y^2}$$

Problem 50. Find the derivative of r and h with respect to one another $f(r,h) = \pi r^2 h$. This equation is the formula for the volume of a cylinder. In other words, the question is how does the volume $f(r,h)$ of the cylinder change when the radius or height is held constant, or when the height is held constant. The way to do this is to hold one constant and change the other:

$f(r,h) = \pi r^2 h$ so hold the radius constant, $\dfrac{df}{dh} = \pi r^2$. So the rate of change of the volume of a cylinder when the radius is held the same but the height is different is $\dfrac{df}{dh} = \pi r^2$.

So, what is the rate of change of the volume of a cylinder when the height is held the same but the radius is changed? The equation is this: $f(r,h) = \pi r^2 h$ so holding the height constant, $\dfrac{df}{fr} = 2\pi rh$. So the rate of change of the volume of the cylinder changes when the height is held constant is $\dfrac{df}{fr} = 2\pi rh.$

Hopefully, in this chapter, you have seen that there are some real-world applications to calculus and that it can be used to determine a variety of complex equations.

CHAPTER 3:
INTEGRAL CALCULUS

Integral calculus is a way of measuring the area and volumes of things that aren't items like spheres, circles, cylinders, or cubes. There are specific calculations that many people know, such as the volume of a cylinder is $\pi r^2 h$. You can calculate the area under a line on a graph such as this one without calculus:

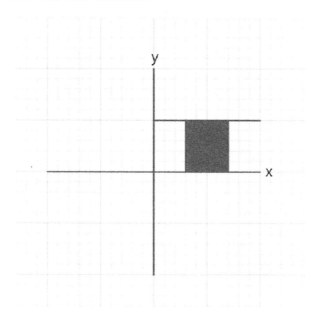

You just have to know that the area under the line between 3 and 7 on the x axis and between 0 and 5 on the y axis to get 5 x 4 = 20 units squared. But what if it's not as simple as that and you have a different graph, such as the one below, you have a bigger problem:

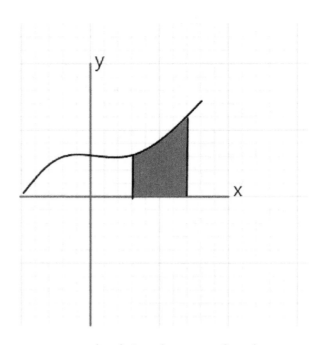

Certainly, you can guess by doing the area of each square and partial square and add it up but this is cumbersome and not accurate. This is simply slicing up columns of Δx over the length of x under the curve. BUT, if you make Δx approach zero, you can determine the area under the curve of a specific equation. The equation of a curve, as you know, is an equation that is not as simple as $y = mx + b$.

Integral calculus is the reverse of differential calculus in that finding the integral of something is the reverse of finding the derivative. If the derivative of x² is 2x, the integral of 2x is x². Of course, in calculus, there are specific notations that are used as notations that look fancy but just represent something you already know:

The integral of 2x is notated like this: $\int 2x\, dx = x^2 + c.$ The dx basically means "the thickness of tiny intervals along the x axis". Why do we have to use the constant "c"? Because when you do the reverse of the integration you are taking the derivative, and you lose the constant because the derivative of a constant is always zero. The constant is lost in the derivative; you have to add it back when evaluating an integral. However, we do not know exactly what the constant is, so we call it an arbitrary constant c.

Rules of Integration

Just as there are rules for derivatives, there are rules for integration. While these can be worked out the hard way, it's best to assume that the mathematicians have already done this and follow the rules as they've been worked out. These are the rules you need to follow:

- The integral of a is $ax + c$
- The integral of x is $\dfrac{x^2}{2} + c$
- The integral of x^2 is $\dfrac{x^3}{3} + c$
- The integral of x is $\ln|x| + c$ (where ln represents the natural logarithm of a number and $|x|$ is the same as saying x has to be positive.
- The integral of e^x is $e^x + c$
- The integral of a^x is $\dfrac{a^x}{\ln(a)} + c$
- The integral of $\ln(x)$ is $x(\ln(x)) - x + c$
- The integral of $\cos(x)$ is $\sin(x) + c$
- The integral of $\sin(x)$ is $-\cos(x) + c$
- The integral of $\sec^2(x)$ is $\tan(x) + c$
- Multiplication by a constant rule: $\int cf(x)dx = c\int f(x)dx$
- Power rule for n not equal to -1: $\int x^n \, dx = \dfrac{x^{n+1}}{n+1} + c$
- Sum rule: $\int (f + g) \, dx = \int f \, dx + \int g \, dx$
- Subtraction rule: $\int (f - g) \, dx = \int f \, dx - \int g \, dx$

Let's do some problems:

Problem 51. Integrate: $\displaystyle\int 3x^2 \, dx$

This is actually an easy problem if you think about what this is (the reverse of the derivative). If you want to play it by the rules, you follow the rule that the integral of $x^2 = \dfrac{x^3}{3} + c$ and the fact that you add the constant, you get $3\left(\dfrac{x^3}{3}\right) + c$ or the integral:

$$\int 3x^2 \, dx = x^3 + c$$

Problem 52. Integrate: $\int 5x^4 \, dx$

First bring out the 5 to get $5\int x^4 \, dx$ and, according to the power rule: $\int x^n \, dx = \dfrac{x^{n+1}}{n+1} + c$, which leads to $\dfrac{5(x^5)}{5} + c$. The answer then is:

$$\int 5x^4 \, dx = x^5 + c$$

Problem 53. Integrate: $\int 5e^x \, dx$

This means solving the exponent rule: $\int 5e^x \, dx = 5\int e^x dx$

$$\int 5e^x \, dx = 5e^x + c$$

Problem 54. Integrate: $\int x^5 dx$

This follows the power rule: $\int x^n \, dx = \dfrac{x^{n+1}}{n+1} + c$ so it's really quite simple:

$$\int x^5 \, dx = \dfrac{x^6}{6} + c$$

Problem 55. Integrate: $\int \sin(x) dx$

This follows basic rules as listed so that the answer is this:

$$\int \sin(x)dx = -\cos(x) + c$$

Problem 56. Integrate: $\int \sec^2(x)dx$

This follows the basic rules as listed so the answer is this:

$$\int \sec^2(x)dx = \tan(x) + c$$

Problem 57. Integrate: $\int \frac{1}{x}dx$

This follows the rule that the integral of $\frac{1}{x} = \ln|x| + c$

$$\int \frac{1}{x}dx = \ln|x| + c$$

Problem 58. Integrate: $\int (5x^4 + \cos(x))\,dx$

Integrate term by term. Using the sum rule (and a few other rules you already know), you get this:

$$\int 5x^4 dx + \int \cos(x)dx$$

$$\int (5x^4 + \cos(x))dx = x^5 + \sin(x) + c$$

Problem 59. Integrate: $\int (3e^x - 5x)dx$

Again, the subtraction rule is this: $\int (3e^x)dx - \int (5x)dx$

This leads to this: $3e^x - 5\frac{x^2}{2} = 3e^x - 2.5x^2 + c$ or $3e^x - \frac{5}{2}x^2 + c$

Problem 60. Integrate: $\int (3x - 1)(3x + 1)dx$

In this case, it is better to multiply the two binomials before integrating to give the integral of $9x^2 - 1$ so the integral becomes $\int (9x^2 - 1)dx$.

The result of the integration is: $9\dfrac{x^3}{3} - x + c = 3x^3 - x + c$

So far, we have been looking at "indefinite" integrals so you could get an idea of what it's like to do an integral. Now we'll look at "definite" integrals, which are far more useful. This involves getting the area under a curve between two definite points, such as a and b along the x-axis. What it looks like is this:

$$\int_{a}^{b} f(x)dx$$

This involves the actual solving of a problem with a real number involved at the end. Let's try it:

Problem 61. Integrate:
$$\int_1^2 2x\,dx$$

Using the rules of integration, you get $x^2 + c$

Evaluating at x = 2, you get $(2)2 + c = 4 + c$

At x = 1, you get $(1)2 + c = 1 + c$

Subtracting the two, you get $(4 + c) - (1 + c) = 3$

The answer is:
$$\int_1^2 2x\,dx = 3$$

Problem 62. Integrate:
$$\int_{-2}^3 3x^2\,dx$$

This problem starts with evaluating the indefinite integral to get
$$3\frac{x^3}{3} = x^3 + c$$

From now on, since we saw that the "+ c" from each integral cancelled in the last problem, we will no longer include a "+ c" when evaluating a definite integral.

Evaluate the result at x = 3, you get $3^3 = 27$

Evaluate the result at x = -2, you get $(-2)^3 = -8$

Subtracting, you get $27 - (-8) = 35$

The answer is:
$$\int_{-2}^3 3x^2\,dx = 35$$

Problem 63. Integrate: $\int_{2}^{4} x^3 \, dx$

First integrate the indefinite integral: $\dfrac{x^4}{4} + c$

Then evaluate the result at x = 4 and x = 2:

For x = 4: $\dfrac{4^4}{4} =$ 64

For x = 2: $\dfrac{2^4}{4} =$ 4

The answer is: $\int_{2}^{4} x^3 \, dx = 60$

Problem 64. Integrate: $\int_{-3}^{3} (x^2 + x + 1) \, dx$

The indefinite integral gives us: $\dfrac{x^3}{3} + \dfrac{x^2}{2} + x + c$

Evaluate that expression for x = 3: $\dfrac{3^3}{3} + \dfrac{(3)^2}{2} + 3$ = 16.5

Evaluate that expression for x = -3: $\dfrac{(-3)^3}{3} + \dfrac{(-3)^2}{2} + (-3)$ = -7.5

Subtracting $16.5 - (-7.5) = 24$

The answer is: $\int_{-3}^{3} (x^2 + x + 1) \, dx = 24$

Problem 65. Integrate: $\displaystyle\int_1^e \frac{1}{x}\,dx$

First integrate the indefinite integral: $\ln|x| + c$

Evaluate for x = e: $\ln|e| = 1$

Evaluate for x = 1: $\ln|1| = 0$

This one isn't fair because you'd have to know that the natural log of e is 1 and the natural log of 1 is 0.

This leads to $1 - 0 = 1$

So, the answer is: $\displaystyle\int_1^e \frac{1}{x}\,dx = 1$

Problem 66. You have a curve that has the equation of $y = -3x^2 + 12$. What is the area under the curve between x = 0 and x = 2?

Basically, this means you are evaluating this integral: $\int_0^2 (-3x^2 + 12)dx$

Start with evaluating the indefinite integral witch gives you:

$(-3)\dfrac{x^3}{3} + 12x$ + c

Evaluate for x = 2: $(-3)\dfrac{2^3}{3} + 12(2)$ = 16

Evaluate for x = 0: $(-3)\dfrac{0^3}{3} + 12(0)$ = 0

The answer is: $\int_0^2 (-3x^2 + 12)dx = 16$

The graph of the area A looks like this:

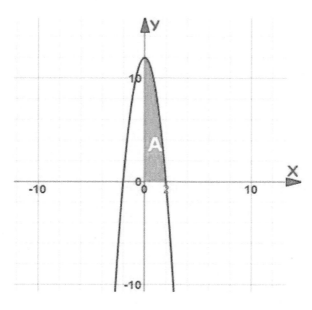

Problem 67. You have a graph of the following equation: $6x^2 + 7 = y$. What is the area under the curve between x = 1 and x = 2?

Evaluate this integral: $\int\limits_{1}^{2} [(6)x^2 + 7]\, dx$

The indefinite integral gives: $(6)\dfrac{x^3}{3} + 7x + c$

Evaluate for x = 2: $(6)\dfrac{2^3}{3} + 7(2) = 30$

Evaluate for x = 1: $(6)\dfrac{1^3}{3} + 7(1) = 9$

Subtracting the two gives: $\int\limits_{1}^{2} [(6)x^2 + 7]\, dx = 21$

Problem 68. What is the area under the curve of $y = 4x^3 + x + 1$ between 0 and 2?

Putting into perspective, we will evaluate the integral:
$\int\limits_{0}^{2} (4x^3 + x + 1)\, dx$

Evaluating the indefinite integral first gives: $4\left(\dfrac{x^4}{4}\right) + \dfrac{x^2}{2} + x + c$

Evaluating for x = 2: $4\left(\dfrac{2^4}{4}\right) + \dfrac{2^2}{2} + 2 = 20$

Evaluating for x = 0: $4\left(\dfrac{0^4}{4}\right) + \dfrac{0^2}{2} + 0 = 0$

Thus, the answer is: $\int\limits_{0}^{2} (4x^3 + x + 1)\, dx = 20$

Problem 69. What is the area beneath the curve of the equation $y = 3x^2 - 2x + 6$ between x = 3 and x = 0?

The integral to evaluate is:
$$\int_0^3 (3x^2 - 2x + 6)\, dx$$

First evaluating the indefinite integral gives: $3\left(\dfrac{x^3}{3}\right) - 2\dfrac{x^2}{2} + 6x + c$

Evaluate for x = 3: $3\left(\dfrac{3^3}{3}\right) - 2\dfrac{3^2}{2} + 6(3) = \quad 36$

Evaluate for x = 0: $3(\dfrac{0^3}{3}) - 2\dfrac{0^2}{2} + 6(0) \quad = 0$

The answer is:
$$\int_0^3 3x^2 - 2x + 6)\, dx = 36$$

Problem 70. What is the area under the curve of a graph that can be described as $y = 6x^3 + x{-}1$ between x=1 and x = 2?

The equation is this:
$$\int_1^2 (6x^3 + x - 1)\, dx$$

Find the indefinite integral: $(6)\dfrac{x^4}{4} + \dfrac{x^2}{2} - x + c$

Evaluate for x = 2: $(6)\dfrac{2^4}{4} + \dfrac{2^2}{2} - 2 \quad = 24$

Evaluate for x = 1: $(6)\dfrac{1^4}{4} + \dfrac{1^2}{2} - 1 = \quad 1$

The answer is:
$$\int_1^2 6x^3 + x - 1\, dx = 23$$

Problem 71. George leaves his home a time $t = 0$ and drives at the rate: $v(t) = 60 - t/2$ miles/hr. (where t is in hours). How far does he drive in 2 hours? Basically, it means that the velocity decreases over time. When evaluating the integral you get the miles George drove in 2 hours.

To get the answer, evaluate the integral: $\int_0^2 \left(60 - \frac{t}{2}\right) dt$

The result of integrating is $60t - \frac{t^2}{4}$

Evaluating at x = 2 you get: $60(2) - \frac{2^2}{4} = 119$

Evaluating at x = 0 you get: $60(0) - \frac{0^2}{4} = 0$

The answer is: $\int_0^2 60 - \frac{t}{2} \, dt = 119$ miles

Problem 72. You turn on a faucet and water runs at a rate of $v(t) = t^3 - \frac{t^2}{2} + 4$ gallons per minute. How many gallons of water are poured out after 2 minutes?

To find the answer to the problem integrate: $\int_0^2 \left(t^3 - \frac{t^2}{2} + 4\right) dt$

The indefinite integral gives: $\frac{t^4}{4} - \frac{t^3}{6} + 4t + c$

Evaluating for 0 gives zero and evaluating for t = 2: $\frac{2^4}{4} - \frac{2^3}{6} + 4(2) = $ 10.67

The answer: $\int_{0}^{2} (t^3 - \frac{t^2}{2} + 4)\, dt = 10.67$ gallons

Problem 73. A cat is climbing a tree and is meowing loudly, more so as she climbs the tree. The volume of the cat's meow, in decibels, is determined by the height as $v(h) = \dfrac{h}{20} + \dfrac{h^2}{3600}$. What is the change in the volume of the cat's meow between 10 feet and 20 feet high?

Answer the problem with this integral: $\displaystyle\int_{10}^{20} \left(\dfrac{h}{20} + \dfrac{h^2}{3600}\right) dx$.

Integrating gives: $\dfrac{h^2}{40} + \dfrac{h^3}{10800}$

Evaluate for h = 20: $\dfrac{20^2}{40} + \dfrac{20^3}{10800} = 10 + .74 = 10.74$

Evaluate for h = 10: $\dfrac{10^2}{40} + \dfrac{10^3}{10800} = 2.59$

The answer is 10.74 − 2.59 = 8.15 decibels

Problem 74. You can paint a wall at a rate of $r(t) = 150 - 4t$ square feet per hour. How much of a wall can you paint in 3 hours?

Solve the problem with this integral: $\displaystyle\int_{0}^{3} (150 - 4t)\, dt$

Integrating gives: $150t - \dfrac{4t^2}{2} + c = 150t - 2t^2 + c$

Evaluate at t = 3 (no need to evaluate at 0 as the answer is 0):
$150(3) - 2(3)^2 = 432$ square feet.

Problem 75. You are painting a fence at a rate of 200 – 4t square feet per hour. How many square feet can you paint in 2 hours?

This integral gives the answer. $\int_0^2 (200 - 4t)\, dt$

The integral is this: $200t - \dfrac{4(t)^2}{2} = 200t - 2t^2$

Evaluate for t = 2 (solving for t = 0 gives 0):

$200(2) - 2(2)^2 = 392\ square\ feet$

CHAPTER 4:
SOLVING SOME TOUGH CALCULUS PROBLEMS: NESTING INTEGRALS, DOUBLE INTEGRALS AND TRIPLE INTEGRALS

This chapter covers some more important integral calculus problems that are based on some things you already know how to do. It gets a little bit more complicated and you'll have to be careful about your equations but these things are important to know about when it comes to understanding calculus. If you still feel like you're on your feet with regard to calculus, we can get started!

Let's first start by doing an integration by substitution or "nesting integrals". This gets tricky because you can't substitute an integration problem as easily as you can substitute a derivative. The only way this works is if the integral is in a specific form. This means that the substitution can only be done if the integral is in a specific way:

$$\int f(g(x))g'(x)\,dx$$. In this problem, you are substituting g(x) into the integral but only if it can be multiplied by its derivative g'(x). This leads to some problems that are easy and some that require some creative organizing of the equation. In such cases, you basically get $$\int f(g(x))g'(x)\,dx = \int f(u)\,du.$$ The $g'(x)$ does not get included in the equation of $$\int f(u)du$$ so it doesn't get added back when you resubstitute the $g(x)$ for u. Let's try an easy one:

Problem 76. Evaluate this indefinite integral: $\int \cos(x^2)2x \, dx$

In this problem, the function $(f(x))$ is $\cos x$, $g(x) = x^2$ is and the derivative $g'(x) = 2x$. Basically, you need to create a whole new integral called $\int f(u)du$, where $u = x^2$. So, what is the solution of $\int \cos(u)du$?

It's $\sin(u) + c$. Now you need to plug in the x^2 for u to get the answer:

$$\int \cos(x^2)2x \, dx = \sin(x^2) + c$$.

This, at least, isn't very messy. Let's try something a little bit more complicated.

Problem 77. Evaluate this indefinite integral: $\int (x+1)^3 \, dx$. You need to rearrange this problem so you can get it into a form where you can multiply the function by its derivative. Well, you know that the derivative of $x+1$ is 1.

So, you can change the integral into this form: $\int (u)^3 \, dx$. Now evaluate the integral in terms of u (Remember, it's an indefinite integral):

$$\int u^3 du = \frac{u^4}{4} + c$$.

Next substitute $(x+1)$ back into the result:

$$\int (x+1)^3 \, dx = \frac{(x+1)^4}{4} + c$$

Problem 78. Evaluate this indefinite integral: $\int (5x + 2)^7 \, dx$

So, how can you get this into the right form of the equation? Use the fact that the derivative of $5x + 2$ is 5 and rewrite the equation: $\frac{1}{5}\int (5x + 2)^7 (5) \, dx$ because the 1/5 and 5 will cancel each other out.

Now evaluate the integral: $\frac{1}{5}\int u^7 \, du = \frac{1}{5}\left(\frac{u^8}{8}\right) + c$

Now plug in the (5x + 2) into the problem: $\int (5x + 2)^7 \, dx = \frac{(5x + 7)^8}{40} + c$

Problem 79. Evaluate this indefinite integral: $\int sinx^2(2x) \, dx$. This integral can be changed to $\int sin(u) du$. Evaluate this integral according to the rules:

$\int sin(u) du = -\cos(u) + c$. Now resubstitute the x² back into the equation to get:

$$\int sinx^2 2x \, dx = -\cos(x^2) + c$$

Problem 80. Evaluate this indefinite integral: $\int \frac{x^2}{x^3 + 1} dx$

What you need to do is to find derivative that works for this: The derivative of x³ +1 is 3x² so you can rewrite the integral as: $\frac{1}{3}\int \frac{3x^2}{x^3 + 1} dx$. Now you can separate out the derivative from its function by writing: $\frac{1}{3}\int \frac{1}{x^3 + 1}(3x^2) \, dx = \frac{1}{3}\int \frac{1}{u} du.$ Looking back on the rules, you know that for the integral of 1/u you get the natural log of the absolute value of u or ln |u|. This means that by substituting everything back, you get:

$\int \frac{x^2}{x^3 + 1} dx = \frac{1}{3}\int \frac{1}{x^3 + 1}(3x^2) \, dx = \frac{1}{3}|x^3 + 1| + c$

Problem 81. Evaluate this indefinite integral: $\int 12(3x+1)^3 \, dx$

What you know is this: the derivative of 3x + 1 is 3. Change the integral to: $\int 4(3x+1)^3(3)dx$. Remember that you can eliminate the three as it "goes with the dx" to make du, so the integral becomes:

$\int 4u^3 \, du = 4 \int u^3 du$. This integral evaluates as $4\dfrac{u^4}{4} + c = u^4 + c$

Substitute it back so you get this: $\int 12(3x+1)^3 \, dx = (3x+1)^4 + c$

Problem 82. Solve this: $\int (2x+3)(x^2+3x)^2 dx$

The first thing you need to do is to determine that the derivative of $x^2 + 3x$ is $2x + 3$. So basically, this means that you can use $2x + 3$ to be the derivative of u, which is $x^2 + 3x$. This gives us the integral: $\int u^2 dx$

This integral becomes: $\dfrac{u^3}{3} + c = \dfrac{1}{3}(x^2+3x)^3 + c$

Problem 83. Solve this: $\int 5sin^4(x)cosx \, dx$

This actually fairly easy because the definitive of $sin(x)$ is $cos(x)$. Make $u = sin(x)$ and the $du = cos(x) \, dx$

This means that the integral becomes: $5\int u^4 du$ so the answer is: (5) $\dfrac{u^5}{5} + c$. Substituting back again, you get: $\int 5sin^4(x)cosx \, dx = cos^5(x) + c$

Problem 85. Solve this: $\int \dfrac{8x}{(1-x^2)^4} dx$

So, what if $u = 1-x^2$ then $du = -2xdx$. This leads to the somewhat messy integral:

$$-4 \int \frac{1}{u^4} du = -4 \int u^{-4} du$$

$$(-4)\left(-\frac{1}{3}\right) u^{-3} + c$$

substituting back in the value of u, you get:

$$\frac{4}{3(1-x^2)^3} + c$$

Now let's look at double integrals: these are similar to double derivatives and have a particular meaning. If the integral of an equation is the area under the curve, the double integral is described as the volume of a specific item that is not a perfect cube, cylinder, or other geometric solid. What if you have a solid like this one?

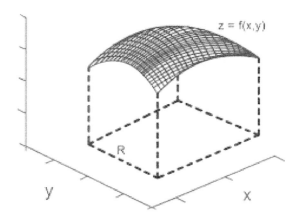

You can use a double integral to find the area under the curved surface. Let's do some double integrals to see how it works:

Problem 86. Evaluate this double integral: $\iint x^2 \, dx dx$

First integrate the initial integral of x^2 to get: $\dfrac{x^3}{3} + c$

Now integrate the second integral: $\int \left(\dfrac{x^3}{3} + c \right) dx$

$$\int \left(\dfrac{x^3}{3} + c \right) dx = \dfrac{x^4}{(3)(4)} + cx + d = \dfrac{x^4}{(12)} + cx + d$$

where c and d are arbitrary constants.

Problem 87. Evaluate this double integral: $\iint (x^2 + 2x + 3) \, dx dx$

First find the single integral: $\int (x^2 + 2x + 3) \, dx = \dfrac{x^3}{3} + 2 \left(\dfrac{x^2}{2} \right) + 3x + c$

Now find the second integral:

$$\int \left(\dfrac{x^3}{3} + x^2 + 3x + c \right) dx = \dfrac{x^4}{12} + \dfrac{x^3}{3} + \dfrac{3x^2}{2} + cx + d$$

Problem 88. Solve this tricky double integral:

$$volume = \int_0^1 \int_1^2 (x^2 + xy^3) \, dy dx$$

First integrate the inner integral by treating x as a constant:

$$\int_1^2 (x^2 + xy^3) \, dy = x^2 y + \dfrac{xy^4}{4}$$

Now evaluate the definite integral using the integration limits:

$$x^2 (2) + \dfrac{x(2)^4}{4} - \left(x^2(1) + \dfrac{x(1)^4}{4} \right) = x^2 + \dfrac{15x}{4}$$

Now integrate the outer integral for x: $\int_0^1 \left(x^2 + \dfrac{15x}{4} \right) dx = \dfrac{x^3}{3} + \dfrac{15x^2}{4 \, 2}$

Evaluating with x = 0 gives zero so evaluate for x = 1 to get: $\dfrac{1}{3} + \dfrac{15}{8} = \dfrac{53}{24}$

Problem 89. Evaluate this double integral:
$$\int_0^2 \int_0^2 (x^2 + xy^2)\,dx\,dy$$

You should know that you can integrate the integrals for x first and then y or integrate the integrals for y first and then x, as it will give you the same answer. Let's integrate the inner integral for x with y as a constant:

$$\int_0^2 (x^2 + xy^2)\,dx = \frac{x^3}{3} + \frac{x^2 y^2}{2}$$

Now evaluate for x = 2 and x = 0 to get the value between these two numbers:

$$\frac{(2)^3}{3} + \frac{(2)^2 y^2}{2} = \frac{8}{3} + 2y^2$$

Now integrate for the outer integral:
$$\int_0^2 \left(\frac{8}{3} + 2y^2\right) dy\,dy = \frac{8}{3}y + 2\frac{y^3}{3}$$

Evaluate the result of integration at y = 2 to yield
$$\frac{8}{3}(2) + 2\frac{(2)^3}{3} = \frac{32}{3}$$

Problem 90. Evaluate this double integral.
$$\int_1^2 \int_0^1 (2x + xy)\,dx\,dy$$

Integrate the inner integral first:
$$\int_0^1 (2x + xy)\,dx = x^2 + \frac{yx^2}{2}$$

Evaluate the result at the integration limits x = 0 and x = 1 (the answer for 0 is 0):
$$1^2 + \frac{y \cdot 1^2}{2} = 1 + \frac{y}{2}$$

Now integrate the outer integral:
$$\int_1^2 \left(1 + \frac{y}{2}\right) dy = y + \frac{y^2}{4}$$

Evaluate the result at the integration limits y = 1 and y = 2: $2 + \dfrac{2^2}{4}$ - $\left(1 + \dfrac{1^2}{4}\right) = \dfrac{7}{4}$

Problem 91. Evaluate this double integral. $\int_0^2 \int_0^1 (x^2 + xy)\, dxdy$

Start with the first (inner) integral: $\int_0^1 (x^2 + xy)\, dx = \dfrac{x^3}{3} + \dfrac{x^2 y}{2} = \dfrac{1}{3} + \dfrac{y}{2}$
(after evaluating at the integration limits).

Now integrate the outer integral: $\int_0^2 \left(\dfrac{1}{3} + \dfrac{y}{2}\right) dy = \dfrac{y}{3} + \dfrac{y^2}{4}$

Evaluate the result at the integration limits: $\dfrac{2}{3} + \dfrac{2^2}{4} - 0 = \dfrac{5}{3}$

Problem 92. Evaluate this double integral. $\int_0^2 \int_0^1 (xy^2 + x^2 y)\, dxdy$

Integrate the inner integral first: $\int_0^1 (xy^2 + x^2 y)\, dx = \dfrac{x^2 y^2}{2} + \dfrac{x^3 y}{3}$

Evaluating at the integration limits (0 and 1), you get: $\dfrac{(1)y^2}{2} + \dfrac{(1)y}{3}$
(Remember that evaluating at zero is zero in these circumstances).

Now integrate the second integral: $\int_0^2 \left(\dfrac{y^2}{2} + \dfrac{y}{3}\right) dy = \dfrac{y^3}{6} + \dfrac{y^2}{6}$

Now evaluating at the limits y = 2 and y = 0, you get:
$\dfrac{2^3}{6} + \dfrac{2^2}{6} = \dfrac{8}{6} + \dfrac{4}{6} = \dfrac{12}{6} = 2$

Problem 93. Evaluate this double integral: $\int_0^1 \int_0^1 (x^2 + 2xy + y^2)dxdy$

Integrate the inner integral first: $\int_0^1 (x^2 + 2xy + y^2)dx = \dfrac{x^3}{3} + \dfrac{2yx^2}{2} + xy^2$

Evaluate at the integration limits x = 0 and x = 1 to get:
$\dfrac{1}{3} + \dfrac{2y(1)}{2} + (1)y^2 = \dfrac{1}{3} + y + y^2$

Now integrate the outer integral: $\int_0^1 \left(\dfrac{1}{3} + y + y^2\right)dy = \dfrac{y}{3} + \dfrac{y^2}{2} + \dfrac{y^3}{3}$

Evaluating at the limits of y = 0 to y = 1, you get: $\dfrac{1}{3} + \dfrac{1}{2} + \dfrac{1}{3} = \dfrac{7}{6}$

Problem 94. Evaluate this double integral: $\int_0^2 \int_0^2 (2x + xy)dxdy$

First, integrate the inner integral: $\int_0^2 (2x + xy)dx = \dfrac{2x^2}{2} + \dfrac{yx^2}{2}$

Evaluate the result at x = 2 to x = 0: $\dfrac{2(2)^2}{2} + \dfrac{y(2)^2}{2} = 4 + 2y$

Now integrate the outer integral: $\int_0^2 (4 + 2y)dy = 4y + \dfrac{2y^2}{2}$

Now evaluate the result at y = 2 and y = 0: $4(2) + \dfrac{2(2)^2}{2} = 12$

Problem 95. In this problem, we will have a triple integral using x, y, and z. This is really no different than evaluating a double integral but it involves three integral problems stacked one on top of the other:

$$\int_0^1 \int_0^1 \int_0^1 (x^2 + y^2 + z^2)dxdydz$$

Okay, so let's get started with the inner integral, keeping the other variables constant:

$$\int_0^1 (x^2 + y^2 + z^2)dx = \frac{x^3}{3} + xy^2 + xz^2$$

Evaluating at the integration limits x = 1 and x = 0, you get $\frac{1}{3} + y^2 + z^2$

Integrate the next integral for y: $\int_0^1 \left(\frac{1}{3} + y^2 + z^2\right)dy = \frac{y}{3} + \frac{y^3}{3} + yz^2$

Evaluating at the integration limits y = 1 and y = 0, you get:
$$\frac{1}{3} + \frac{1}{3} + z^2 = \frac{2}{3} + z^2$$

Finally, evaluating the outer integral, you get: $\int_0^1 \left(\frac{2}{3} + z^2\right)dz = \frac{2z}{3} + \frac{z^3}{3}$

Finally, when you evaluate the results at the integration limits of z from 0 to 1: $\frac{2}{3} + \frac{1^3}{3} = 1$ Wasn't that simple in the end?

Problem 96. Evaluate this triple integral: $\displaystyle\int_0^2 \int_0^1 \int_0^1 (xy^2 - xy + yz^2)dxdydz$

Integrate the first integral for x: $\displaystyle\int_0^1 (xy^2 - xy + yz^2)dx = \frac{x^2 y^2}{2} - \frac{yx^2}{2} + xyz^2$

Now, evaluate the results for x = 1 and x = 0 (knowing the answer for x = 0 will be zero): $\displaystyle\frac{y^2}{2} - \frac{y}{2} + yz^2$

Now, integrate the second integral for y:

$$\int_0^1 \left(\frac{y^2}{2} - \frac{y}{2} + yz^2\right) dy = \frac{y^3}{6} - \frac{y^2}{4} + \frac{y^2 z^2}{2}$$

Evaluate the results at the integration limits y = 1 and y = 0 to get:

$$\frac{1}{6} - \frac{1}{4} + \frac{z^2}{2} = -\frac{1}{12} + \frac{z^2}{2}$$

Now integrate the third integral: $\displaystyle\int_0^2 \left(-\frac{1}{12} + \frac{z^2}{2}\right)dz = -\frac{z}{12} + \frac{z^3}{6}$

Now evaluate the results at the integration limits for z from 0 to 2:

$$-\frac{2}{12} + \frac{2^3}{6} = \frac{7}{6}$$

Problem 97. Evaluate this triple integral:

$$\int_0^1 \int_0^1 \int_0^1 (2x^2 + 2xy + 2z^2)\,dx\,dy\,dz$$

Integrate the inner integral for x:

$$\int_0^1 (2x^2 + 2xy + 2z^2)\,dx = \frac{2x^3}{3} + \frac{2yx^2}{2} + 2xz^2$$

Evaluate the results at the integration limits from x = 0 to x = 1 to get:

$$\frac{2(1)^3}{3} + \frac{2y(1)^2}{2} + 2(1)z^2 = \frac{2}{3} + y + 2z^2$$

Now integrate the second integral for y:

$$\int_0^1 \left(\frac{2}{3} + y + 2z^2\right) dy = \frac{2y}{3} + \frac{y^2}{2} + 2yz^2$$

Evaluate these results at the integration limits from y = 0 to y = 1 to get:

$$\frac{2}{3} + \frac{1}{2} + 2z^2 = \frac{7}{6} + 2z^2$$

Now integrate the outer integral for z:

$$\int_0^1 \left(\frac{7}{6} + 2z^2\right) dz = \frac{7z}{6} + \frac{2z^3}{3}$$

Evaluate at the integration limits from z = 0 to z = 1 to get:

$$\frac{7}{6} + \frac{2}{3} = \frac{11}{6}$$

Problem 98. Evaluate this triple integral. $\int_0^2 \int_0^1 \int_0^1 (x^2y + xz + yz^2)dxdydz$

Integrate the first (inner) integral for x:

$$\int_0^1 (x^2y + xz + yz^2)dx = \frac{x^3y}{3} + \frac{zx^2}{2} + xyz^2$$

Evaluate the results at the integration limits for x = 1 and x = 0 to get:

$$\frac{y}{3} + \frac{z}{2} + yz^2$$

Now integrate the second integral for y:

$$\int_0^1 \left(\frac{y}{3} + \frac{z}{2} + yz^2\right)dy = \frac{y^2}{6} + \frac{zy}{2} + \frac{y^2z^2}{2}$$

Evaluate at the integration limits y = 1 and y = 0: $\frac{1}{6} + \frac{z}{2} + \frac{z^2}{2}$

Now integrate the outer integral for z: $\int_0^2 \left(\frac{1}{6} + \frac{z}{2} + \frac{z^2}{2}\right)dz = \frac{z}{6} + \frac{z^2}{4} + \frac{z^3}{6}$

Now evaluate at the integration limits, z = 2 and z = 0: $\frac{2}{6} + \frac{4}{4} + \frac{8}{6} = \frac{8}{3}$

Problem 99. Evaluate this triple integral: $\int_0^1 \int_0^1 \int_0^1 (xy+z)\,dx\,dy\,dz$

First integrate the inner integral for x: $\int_0^1 (xy+z)\,dx = \dfrac{yx^2}{2} + xz$

Evaluate these results for x = 1 and x = 0 to get: $\dfrac{y}{2} + z$

Now integrate the second integral for y: $\int_0^1 \left(\dfrac{y}{2} + z\right) dy = \dfrac{y^2}{4} + zy$

Evaluating at the integration limits from y = 0 to y = 1: $\dfrac{1}{4} + z$

Now integrate the outer integral for z: $\int_0^1 \left(\dfrac{1}{4} + z\right) dz = \dfrac{z}{4} + \dfrac{z^2}{2}$

Evaluate these results at the integration limits from z = 0 to z = 1:
$\dfrac{1}{4} + \dfrac{1}{2} = \dfrac{3}{4}$

Problem 100. Integrate this final triple integral:

$$\int_0^2 \int_0^1 \int_0^1 (xy + yz + xz)\,dx\,dy\,dz$$

First integrate the inner integral for x:

$$\int_0^1 (xy + yz + xz)\,dx = \frac{yx^2}{2} + xyz + \frac{zx^2}{2}$$

Evaluate these results at the limits from x = 0 to x = 1 to get: $\frac{y}{2} + yz + \frac{z}{2}$

Next integrate the middle integral for y:

$$\int_0^1 \left(\frac{y}{2} + yz + \frac{z}{2}\right) dy = \frac{y^2}{4} + \frac{zy^2}{2} + \frac{zy}{2}$$

Evaluate the results at the integration limits from y = 0 to y = 1:
$$\frac{1}{4} + \frac{z}{2} + \frac{z}{2} = \frac{1}{4} + z$$

Now integrate the outer integral for z: $\int_0^2 \left(\frac{1}{4} + z\right) dz = \frac{z}{4} + \frac{z^2}{2}$

Evaluate the results for the integration limits from z = 0 to z = 2, you get this: $\frac{2}{4} + \frac{2^2}{2} = \frac{5}{2}$

Hopefully, after a hundred problems involving a wide variety of calculus problems, including differential and integral calculus, as well as some word problems, you realize that calculus does have some real-world applications. Aside from the technicalities of manipulating equations, calculus isn't really as difficult as it might sound like and you, too, can pass any calculus class you might encounter.

We hope you enjoyed this book! If so, can you leave a review on the Amazon book page? It would be greatly appreciated!

If you have any suggestions on ways to improve this book, please contact us at: support@mathwizo.com